Using the
EASB Crossbow CNC Plasma Cutter with FastCAM and Open-Source Graphics Software

First Edition

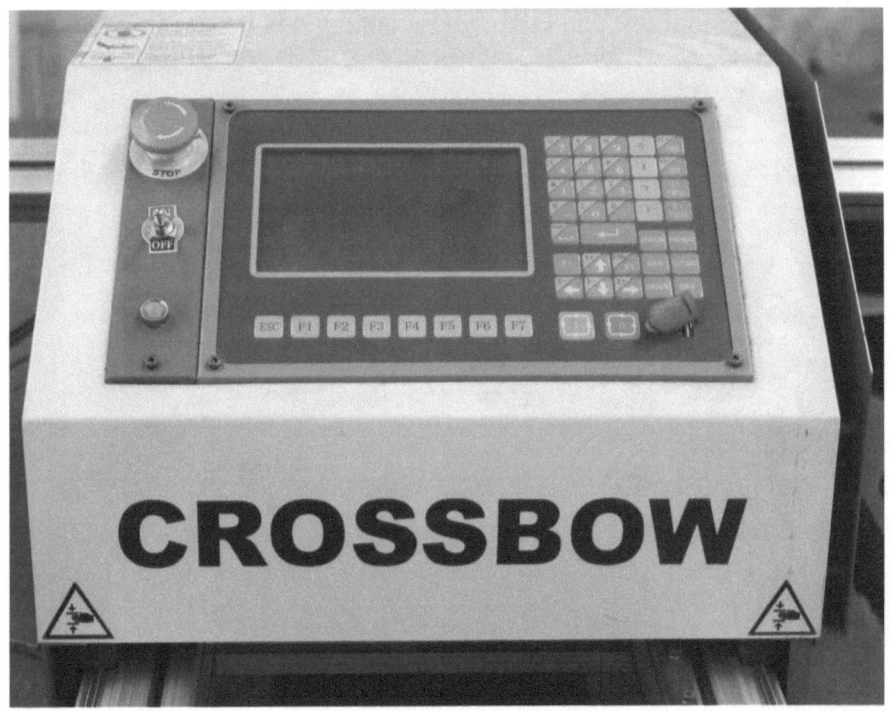

By

J. Burton Browning, Ed.D. and Donald McKeithan

Author contact information:

jbbrowning@jbbrowning.com
http://www.jbbrowning.com

donaldmckeithan89@gmail.com

Table of Contents

Chapter 1
Know your device

Safety first!!!
Of course it stands to reason that before reading any more in this manual, the user will have reviewed the ESAB Crossbow(c) instruction manual, as well as the plasma cutter manual.

Directions serve to orient the user in proper operation of the device, and ensure safety for both the operator and any other people in the shop. Important note: Safety in both device operations and facility rules are your top priority. Do not take shortcuts to save time when you know the safest procedure. One second of miscalculation or unsafe behavior can lead to a lifetime of suffering or worse death. Be the safest person in the shop and set an example for others....they may thank you one day!

Review this text in its entirety before using the device. The steps listed here are to keep you safe, save you aggravation, and increase your success in using this machine.

Clean air and air pressure

You will need clean air, dry air, and oil-free air. A compressor without a water separator and that is not listed as "oil free" will cause poor cuts and will ruin the tips of the plasma cutter in short order.

This is critical! Since the process behind a plasma cutter is to melt metal and blow it away from the surface via air pressure. Moisture or oil in the air line will accelerate wear on the cutting tips, create poor cuts, and will disable the machine.

Get some wind on it! For larger or more time-consuming cuts, you should ensure that you have reserve air pressure. A "pan cake" style air compressor will not suffice for the Crossbow. A tank with approximately 25 gallons of air or more is required.

For success you should:

1) Check that air is filtered/water trapped and oil free.
2) Check that moisture trap is drained each time before using the device.
3) Check that you have proper pressure, as measured at the plasma cutter, NOT at the air compressor. Pressure will drop depending on distance of your piping system for air, so your only concern is if the pressure is correct at the plasma cutter. For thinner metal, you can lower the air pressure at the plasma cutter. For thicker raise it. For sheet metal (thin) at lower amperages, say 30 amps, you can have 50 or less PSI at the plasma cutter unit. For thicker, such as 1/4 plate, try for 80-90 at the cutter. Note this is not on the compressor, the compressor will be higher PSI and you should adjust pressure via the front controls on the ESAB plasma cutter.
4) As compressors vary in capacity, check often that the pressure is correct often while cutting since if not you may have to wait a few minutes to build up reserve pressure in the tank, which will create problems and ruin material.

Basic components

The plasma cutter is only part of this CNC (Computer Numeric Control) cutting system. In fact you might work with a hand plasma cutter only for some tasks. However, with the ESAB Crossbow system, the plasma cutter works as part of the entire automated cutting system, in the same way the air compressor works with it, the cutting table, etc. You should become familiar with the entire system operation before using the ESAB Crossbow CNC system.

Figure 1.0 ESAB *Powercut 900* plasma cutter.

In fact, many plasma cutters do not support remote operations as required by the Crossbow, so make sure that the unit you have will function with the Crossbow. The ESAB 900 is a unit which supports remote control.

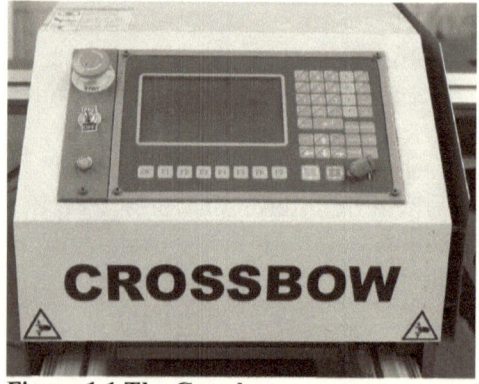

Figure 1.1 The Crossbow computer control unit.

Figure 1.2 The torch head and motorized lift.

Hands-on: Learning about the machine

Using the links provided, watch the short videos and answer the questions listed. The main URL for ESAB and other welding systems is here: http://training.victortechnologies.com/ however use the specific links listed below:

1) Create a free account via the URL below:
http://training.victortechnologies.com/index.php?p=signup

2) Login, then browse to the following URL's and take (and pass!) the quiz at the end. Only the first URL has a quiz.

http://training.victortechnologies.com/index.php?p=asset&asset_id=34

http://training.victortechnologies.com/index.php?p=asset&asset_id=159

4) Navigate to "My Account" and print the successfully completed CNC Basics activity and ESAB Crossbow activity (watched) and give to your professor.

For fun you can search *YouTube* for many examples of the Crossbow, such as this one showing a gear being cut.
https://www.youtube.com/watch?v=yhancflWnP0
Now that you have a good basic overview, let's consider some other important items.

Basic design theory
For all effective purposes the ESAB Crossbow can be used in three ways, all of which are quite simple and straightforward.

   1) It can be used with only the built-in drawings contained in the unit itself.

   2) It can be used with designs made with FastCAM or similar software.

   3) The unit can be used with designs from clipart, prepared in other software applications, and imported for final processing with FastCAM.

All three methods are useful, and provide solutions to specific needs. You will learn to use all three methods via this text.

Correct tip and amperage

There are tips designed for different amperage ranges. For example, some tips are rated (which is on the tip) for no more than 50 amps. This is a tip for no more than 1/4 inch metal. If you use more than 50 amps from the plasma cutter, you will shorten the life of the tip, given that it was a 50 amp tip. *Amperage will be adjusted based on tip type being used.

Figure 1.3   50 amp tip.

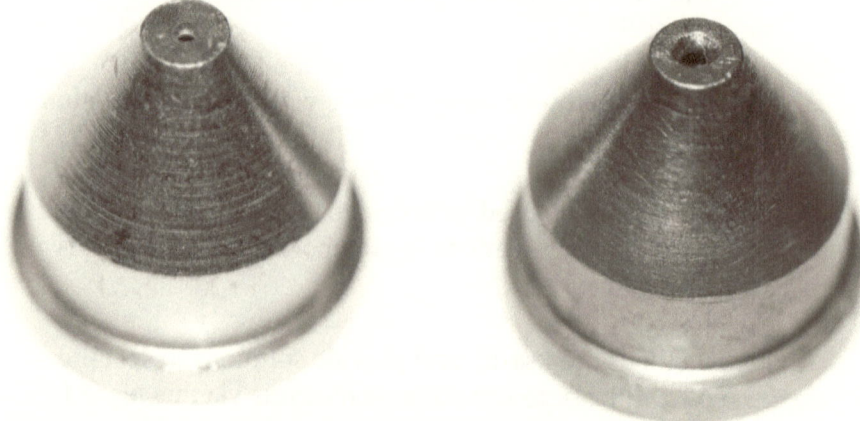

Figure 1.4   New tip on left, worn tip on right.

Figure 1.5 The wear effects the flow of air, causing poor cuts. Note hole is "rounded" out and not a tight clean circle.

There are tips in the up to 30 amp range, up to 50 amps, and on up to 90 amps. A good rule of thumb would be to run your amperage only close to the max rating. E.g. if a 50 amp tip, set the plasma cutter to no higher than about 47 amps. This can extend your tip life. Note that tips and other related parts (baffle chamber, etc) are part of a unit and correctly rated tips should be installed with correctly sized baffles.

Clean metal and good negative ground
You must have a good clean ground for the negative ground clamp on the work you are to cut. Even more importantly the metal should be rust free, if not you will need to sand or grind before cutting. This will severely minimize tip life (of the cutter) regardless of if you are using a CNC cutter or hand plasma cutter.

Your work surface must also be level. E.g. the metal should not be bowed or flex when placed in the cutting area. If so you will find that the Crossbow(c) constantly auto height adjusts (it will appear that it bounces up and down on the cutting surface) trying to find a good distance for ground, which it bases on electrically conductivity, which is effected by cleanliness of the metal surface and the distance between the tip and cutting surface.

Changing Tips and Torch Construction

What follows are instructions to take the torch head apart, and change tip size (for different amperage ranges), cleaning, etc. You would want to change the tip size, based on the thickness of the material you are cutting (to match the amperage range). Also if you see rough cuts or the unit is not functioning properly, it could well be that cleaning or replacing some of the tip parts will be needed.

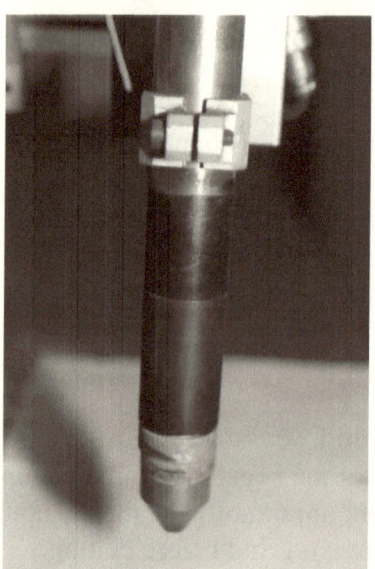

Figure 1.6 Complete torch head assembled.

1) Make sure air is off, Crossbow is off, and plasma cutter is off. Disconnect red ground from head via the spade connector.

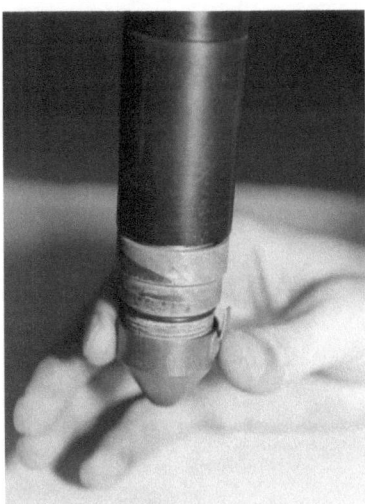

Figure 1.7 Unscrewing cup from torch.

2) Unscrew the cup from the torch. You will want the ESAB o-ring lubricant available for re-assembly. Do not overly apply lube when reassembling, but make sure o-ring has it evenly applied around all rubber surfaces. This will help prevent air leaks.

Figure 1.8 Unscrewing next part of torch, containing baffle, tip, etc.

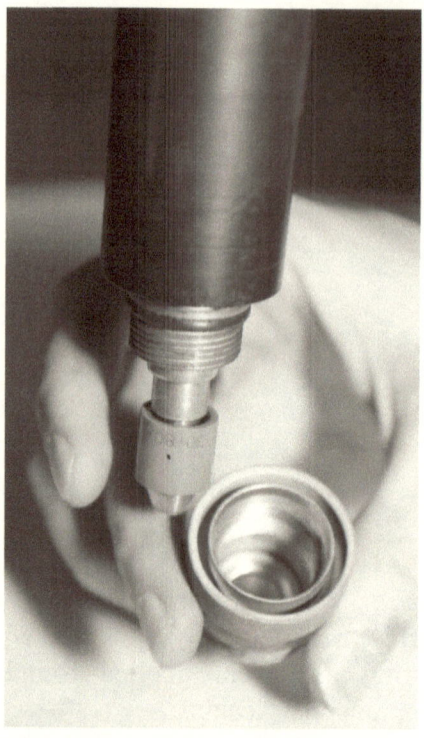

3)

Figure 1.9 Inside showing baffle (on torch) in proper orientation with amperage range at top.

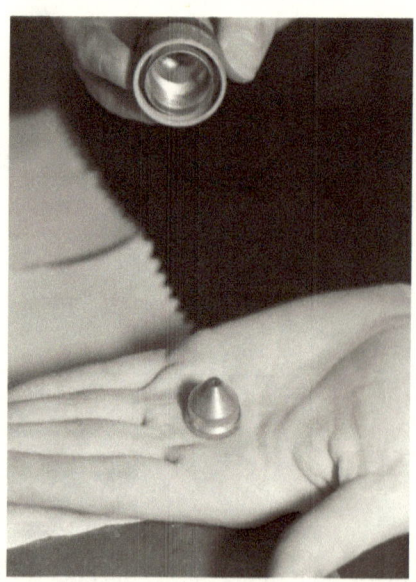

4)

Figure 2.0 Tip (varies with amperage range).

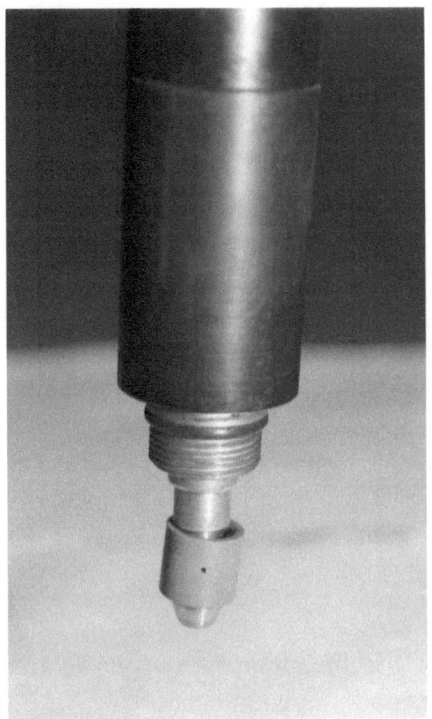

5)

Figure 2.1 Baffle will "wobble" if put on upside down.

Figure 2.1 shows proper orientation, with amperage printed at the top. Note you can partially see numbers on the left of the baffle in figure 2.1

Also note that the o-ring should have a very small amount of o-ring lube applied during reassembly to prevent binding. Use only EASB approved grease. If this is not done it will be tough to dissemble and change tips.

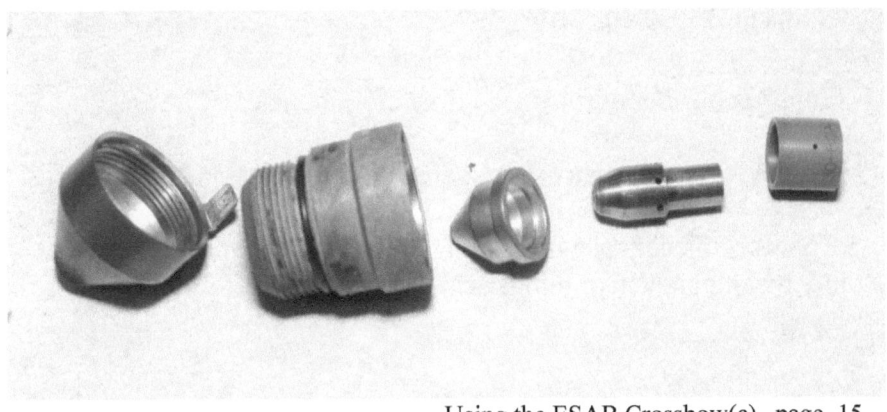

Figure 2.2 All pieces in order of assembly.

6)

Figure 2.3 Baffle with amperage range.

7) Use the appropriate air baffle for the amperage range you are cutting with for long service life and precision cuts.

Safety first, this machine is automated!
It is important to note that this is automated equipment, and as such it will perform the tasks you program it to do, and will not stop if a person or part is in a dangerous area near the cutter. So, you should announce that the cutter is in operation to all nearby, ensure that the cutting arm will not contact people or parts nearby, and most importantly you (the operator) should stand near the emergency stop button and pay close attention to the entire cutting operation until completed.

Things to consider during CNC operation include:

> Do not leave the control panel area until the plasma cutter is off.

> The work piece/table are energized as is the torch tip...note what other metal conductive things are in contact.

> ESD (Electro Static Discharge) is of high concern around any welding equipment, if you have electronic devices, a pacemaker, etc. you should be shielded or away from this device.

> Wear your safety helmet, gloves, have water in the cutting table tank, and have proper spark shields or other protective devices to contain errant sparks while cutting.

You are the one responsible for safety! Know that the area is clear, people in the work area are notified, and you are ready to hit the emergency stop should a problem arise.

Reinforce your skills
Try the following to reinforce what you have learned.

1) Take the torch tip apart and reassemble in proper order.

2) Write a "how-to" or create a presentation of how you would teach another employee about how to work with the Crossbow.

3) Research the cost of "consumables" for the Crossbow.

4) Review the instruction manual for the Powercut plasma cutter (or plasma machine you are using).

5) Review the Crossbow manual provided by ESAB.

6) Research how to drain and clean the compressor and plasma cutter water traps.

Notes:

_____
_____
_____
_____
_____
_____
_____
_____
_____
_____
_____
_____
_____

Chapter 2
Using built-in designs on the Crossbow

The Crossbow has many built-in designs which merely need to be selected and sized to be used. No other computer work is required. Items such as flanges for pipes, angle parts, and gears are pre-drawn for you. Follow along to see how easy it is to cut a part from built-in designs on the Crossbow.

Hands-on: Cutting a flange with the Crossbow

1) Turn on the air compressor and make sure there is water in the cutting table.
2) Make sure the plasma cutter if off.
3) Turn the Crossbow on using the back power switch.
4) Press F6 Library.
5) Use the arrow keys or the right of the keypad to move down to 4-Hole flange.

Figure 2.1

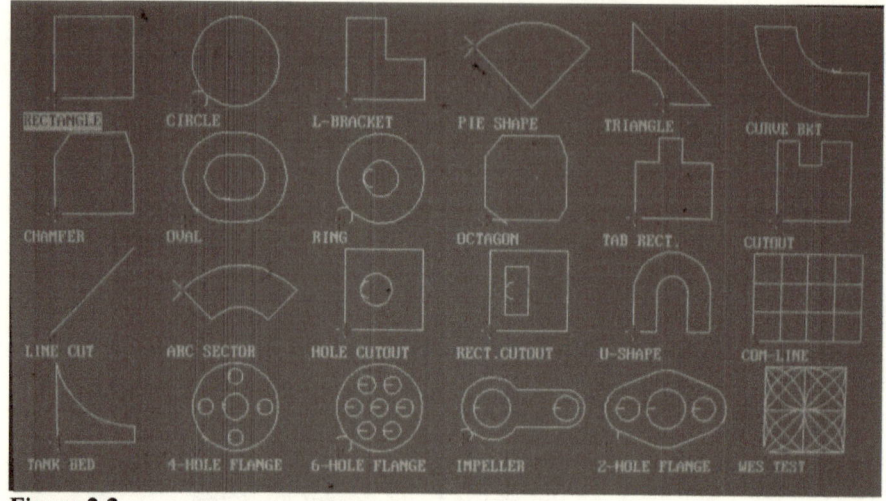

Figure 2.2

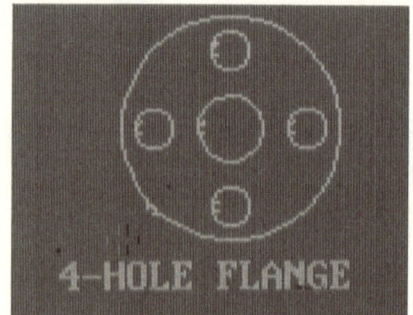

Figure 2.3

6) Press the Enter key (large arrow point left). You should see the flange.

7) On the upper right are settings for the flange size. Use the numeric keypad to change the size in inches to:

Outer R 8.0   (outer radius)
Inner R 2.0   (Inner radius)
Hole R .5   (hole radius)
Hole AX 5.0
Lead-in .5
Salida .5

* Arrow key up if you made a mistake.
F6 apply. You will see the design.

8) To see a change, arrow down to Hole AX and change it to 6.0, Enter, and F6 to apply.

* Note flange bolt holes are closer to the edge now.

9) Manually position cutting head where to start cutting by hand. You should of course have your work piece in place on the table with ground clamp on clean ground.

10) Press Esc.

11) Turn on the motor control for the Crossbow with the front power button.

12) Press F3 Edit.

13) Press F3 Save.

14) Where you see File: (middle bottom of screen) use shift to access the letters on the keypad and name the file, with perhaps initials or something but keep the *.nc* extension. This stands for *Numeric Control.*

15) Press Esc.

16) Press F1 Auto mode.

17) Press F4 Preview cut points will be shown.

18) Press F6 More.

19) Press F1 Outline -- Start from here and press Enter.

20) On upper right, press X to enable Dry Run.

21) Press F4 Preview.

22) Press the green button on the lower right.

23) If you are happy with the dry run, turn off the front of the Crossbow, reset the cut head to position, turn it back on, turn on the plasma cutter, make sure ground clamp is on, check amps and air pressure on plasma cutter, and press the X to remove Dry run. Put on welding helmet and press green button again to start your plasma cut. When completed, turn off plasma cutter before removing work.

Figure 2.4 Flange (How in the world did Mia and Sienna get in this picture?)

Figure 2.5 Flange welded to pipe system.

Reinforce your skills
Try the following to reinforce what you have learned.

1) Cut an L bracket based on data provided by your professor.

2) Cut a circle based on data provided by your professor.

3) Cut a two-hole flange based on data provided by your professor.

4) Cut a four-hole flange based on data provided by your professor.

5) Cut a curved bracket based on data provided by your professor.

6) Chose a template to cut out. Explain which sizes you chose and why.

Notes:

_____

_____

_____

_____

_____

_____

_____

_____

_____

_____

_____

Chapter 3
Using FastCAM(c) software

As there are different versions of FastCAM available, these steps are similar for many versions of FastCAM(c) for Windows. Other versions will be similar, but as the old joke goes "your mileage may vary" so keep this in mind when changing versions.

FastCAM is Computer Aided Manufacturing (CAM) software which allows you to create designs for use on devices, such as the Crossbow CNC Plasma cutter, or import drawings which can then be "cleaned up" for use in the Crossbow, or other devices. First we will use FastCAM to import and convert a vector-based DXF drawing file for use in the Crossbow. Since FastCAM requires a hardware serial key to run, using other computers, with free software to create your DXF files and only using FastCAM to covert to NC code will allow many people to work at the same time on cutting projects. First, let's learn the most productive basics of FastCAM.

Use an existing DXF file with FastCAM
Locate a DXF format file. You can create it or search the web. A good site to find it on the web (Internet) for free and low-cost is: https://freedxf.com/product-category/free/ Search and find a nice design you like. You can find other sites on the web as well with free DXF files for CNC use. Start with a simple design first!

Download and save the file to a location you can access, such as a pen/flash drive or the desktop. Next, use the following steps to load it and create your NC (Numeric Control) code for the Crossbow. For the purposes of this example, a deer DXF file was used, your process however will be similar regardless of the file.

1. Start FastCAM (making sure the hardware key is inserted) and click File, followed by DXF Restore.

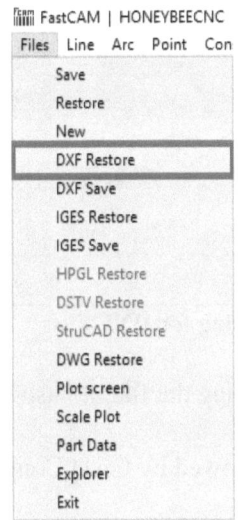

Figure 3.1 DXF Restore option in Files menu.

2. On the DXF Options pop-up, ensure the DXF File Units are correct. If you've been using Metric, stick to Metric. If you're Imperial, standard, etc, choose Inch.

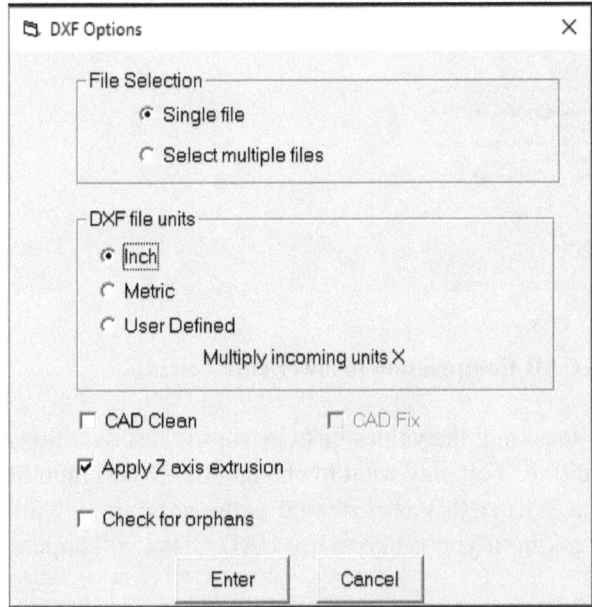

Figure 3.2 DXF Options sample

3. Navigate to your ".dxf" file, click it, and click the select button.

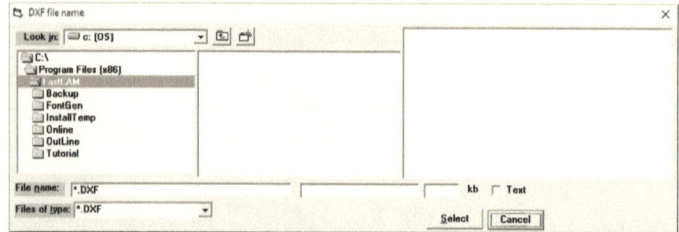

Figure 3.3 Browsing for DXF Files.

4. This should display the file in FastCAM.

5. Click Erase, followed by CAD Compress.

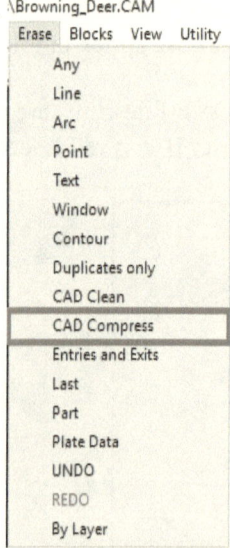

Figure 3.4 CAD Compression to lower entity counts.

Typically speaking, the values FastCAM provides over this next section are fine. You may want to change the, "Maximum Shape Divergence." Keep this as close to 0 as the program will allow.

6. A pop-up asking if you'd like to use CAD Clean will appear, press "no".

7. The next pop-up will ask you what the, "Maximum line length to round" should be. I recommend leaving this value alone. Press Enter.

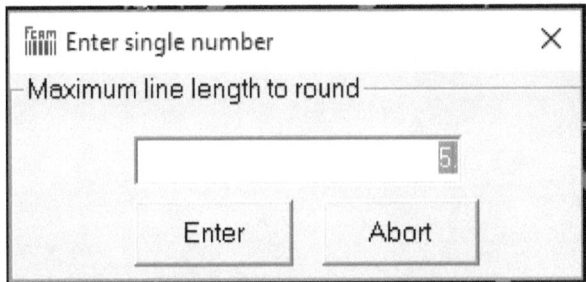

Figure 3.5 Default maximum rounding length.

8. Another pop-up will appear for the "Maximum Shape Divergence." Try to keep this as close to 0 as the program will allow.

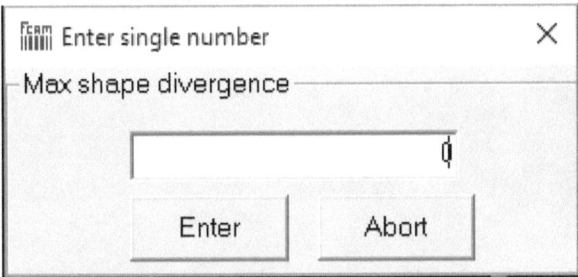

Figure 3.6 Zero shape divergence.

9. The final pop-up will ask if you would like Fastcam to, "Compress arcs." Click yes.
10. This step is where the program may or may not shine. Click Program Path, followed by FastPATH.

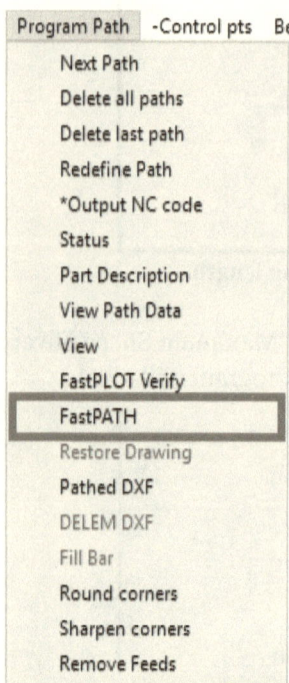

Figure 3.7 Run FastPATH.

11. A menu will pop up. Click Start FastPATH.

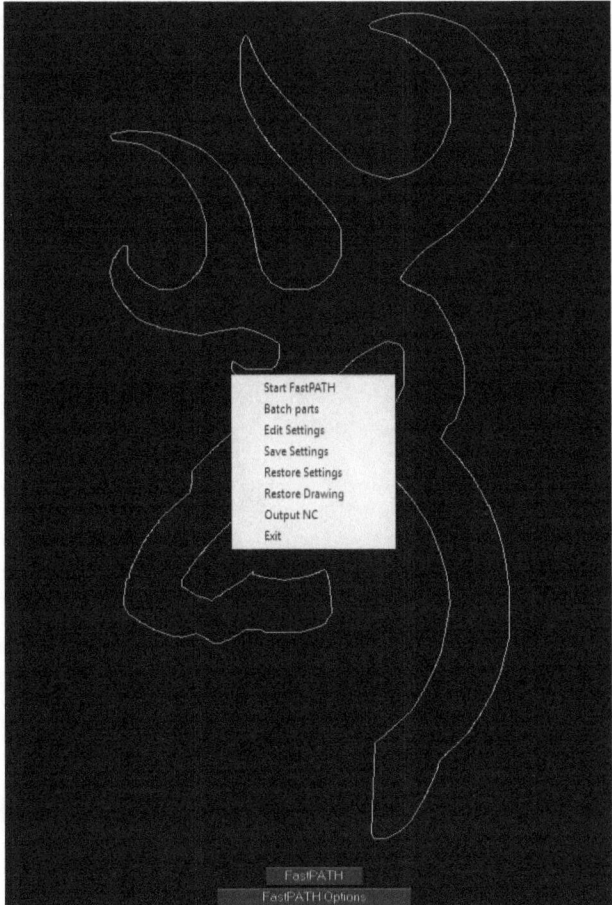

Figure 3.8 Starting FastPATH.

12. When asked if you would like to output the NC code, click "yes".
13. Name and Save your NC code. The name must end with ".txt"
14. Click save.
15. A pop-up will ask if the code should run rapidly at the start, click no.
16. Next, you will be asked if you would like to leave FastCAM and verify the file created, click "yes".
17. This will bring up the image with arrows following the lines of the image. As long as all lines are white, the file will work properly. If there are several red, dotted lines, attempt the next step #18. Otherwise, open the file with a text editor, such as Wordpad, Word, or Notepad, delete any comments at the beginning of the file (there may be 3-6 lines at the start of the file that do not start with a

number and have file name, etc. info.) Delete these and save the file
as text (keep current format) then take the file to the Crossbow for
cutting, you are ready! Steps for loading the file are in chapter 4.

Figure 3.9 Clean cut file.

18. Return to FastCAM.
19. Click Erase, then UNDO.

20. This step is tedious and time consuming, but required if FastPATH didn't work properly. Click Program Path, followed by Next Path.

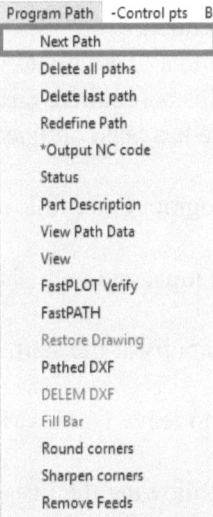

Figure 3.10 Manual tracing.

21. Select CUTTING from the pop-up menu.
22. Select the side of your line you'd like the cut to be made. This depends on the direction you choose to outline the image. If you're outlining in a clockwise motion, choose left to cut on the outside of the line. It is a good idea to outline in a clockwise direction. After selecting your cut side, click anywhere on the image once. This will put an arrow on the image. Moving from one side to the other will change the direction of the arrow. This is the cut direction. Click again to activate the outline. When doing this imagine you are the cutter, moving logically along a path, so you are "teaching" the machine in these steps how to perform the cuts.

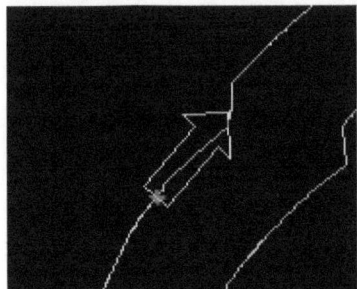

Figure 3.11 Cut orientation arrow.

23. A pop-up will ask you if you want to add an "Entry." This is how you would place a cut-in point. If this is required, click "yes". Cut-ins allow for cleaner cuts and are recommended.
24. Another pop-up will ask if you would like to add an "Exit." This is how you would place a cut-out point. If this is required, click "yes".
25. Repeat steps 20 – 24 until the whole image has been outlined and no blue is showing through anymore.
26. Once the full outline is complete, click Program Path, followed by, "*Output NC Code."
27. Name and Save your NC code. The name must end with ,".txt".
28. Click save.
29. A pop-up will ask if the code should run rapidly at the start, click "no".
30. Next, you will be asked if you would like to leave FastCAM and verify the file created, click "yes".
31. This will bring up the image with arrows following the lines of the image. As long as all lines are white, the file will work properly. If there are several red, dotted lines, attempt again steps 18 - 31. Otherwise, take the file to your machine for cutting after removing any comments in the text file with a text editor such as Wordpad.

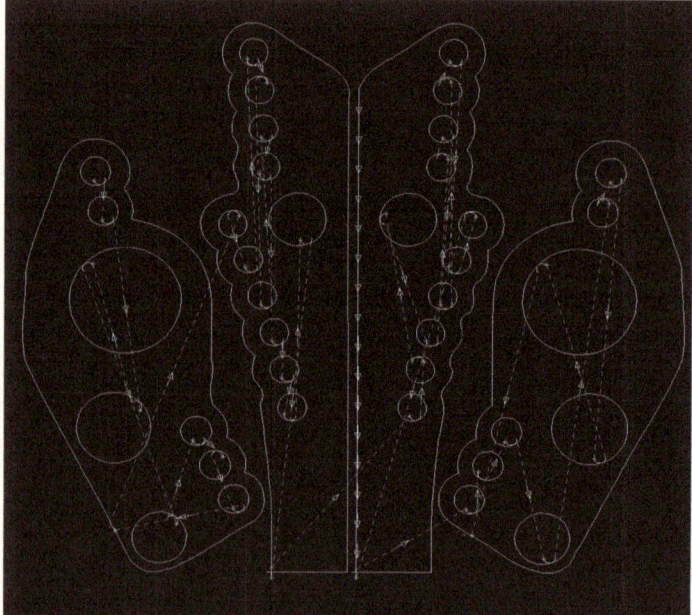

Figure 3.12 Complex bracket file successfully traced.

Reinforce your skills

Try the following to reinforce what you have learned.

1) Locate a *.dxf file on the Internet of an animal or sea creature, such as a starfish. Create a file which you can use on the Crossbow. When finished, cut out the design.

2) Locate a *.dxf file on the Internet of a logo or word or letter. Create a file which you can use on the Crossbow. When finished, cut out the design.

3) Locate a *.dxf file on the Internet of a car, boat, or similar. Create a file which you can use on the Crossbow. When finished, cut out the design.

4) Create your own graphic design, and export as a *.dxf formatted file. Create a file which can be used on the Crossbow.

5) Your supervisor needs cutouts for a shooting target system. You will need various animals for it, such as a deer, lion, gopher, etc. Design these files for use with the Crossbow.

Links:

http://www.readytocut.com/community/
http://vectorink.com/
http://www.cerebralmeltdown.com/cncstuff/filesites/index.htm
http://www.craftsmanspace.com/free-patterns/builders-hardware-patterns.html
https://www.diyweldingplans.com/collections/all
http://mydxf.blogspot.com/
https://grabcad.com/library/halloween-pumpkin-cat-1

Notes:

_____
_____
_____
_____
_____
_____

Chapter 4
Using Gimp, LibreCAD and Inkscape software

The general process is that you will either create your own drawing to import into FastCAM or you will find clipart on the Internet to ultimately import into FastCAM. We say *ultimately* since there well may be cleanup work required on the file before you import it to Gimp and Inkscape.

Installing Gimp
Before you start to use the image processing paint program Gimp, you will need to install it to your computer. There are a few things that will ease this installation process, namely you should have an Internet connection, and you must have administrator rights to install software.

So, a typical lab computer would not work for this unless the computer was either unlocked or had the software already installed. If you have your own computer, you will (if using M.S. Windows) need to right click on the setup/installer and select "Run as administrator" to escalate your privileges to Administrator (and this assumes you are logging in from an administrator privileged account. If you are the sole user of say a personal laptop you may well be logging in with enough rights to perform this task.

Hands-on: Install Gimp
To download and install the software for a M.S. Windows installation, perform the following:
1) Visit http://www.gimp.org/
You will find much information here as well as tutorials, although certainly YouTube is another location for video tutorials (some better than others!).
2) Click on the download link at the top which will bring you to the current download page. http://www.gimp.org/downloads/
Click "Download Gimp Directly".
3) Depending on your browser, you should see the download status. Click to open the folder with the file, then *right click on the file* and select "Run as administrator".

It is important to install in this manner to avoid any install restrictions. If you cannot open the folder, go to My Computer, select the Downloads folder, and in there you will find the Gimp installer (which you will right click on). Install with all defaults.

Hands-on: Install Inkscape
You will also need a vector-based drawing package. Inkscape is free and very powerful. Follow the steps to install it.
1) Go to https://inkscape.org/en/
2) Select the Download menu, and Windows (if using Windows). If using a Macintosh you will also need XQuartz as well.
3) Click the green arrow for the correct version OS you have (probably 64 ibt).
4) Once downloaded, open the folder containing the installer, and right click, select Run as Administrator, and install it. *Note you will need to have admin rights on the machine you are installing to.

It is beyond the scope of this text to teach you how to make a drawing in Gimp (or any other image package) however there is no reason why you cannot make your own creative drawings in Gimp and not worry about using pre-drawn clipart. Certainly Gimp, Paint.NET, and other applications will do a better job than the built-in paint software with Windows. If you are creative, try your hand at drawing on your own and just save the file as DXF format.

Hands-on: Preparing a drawing in Gimp based on downloaded clipart
There is no reason why you could not use Gimp to draw something, then save it for use in FastCAM. However, you may want to start by using some of the links at the end of this chapter to find a clipart image and save it for cleanup and later use in FastCAM.

Using the links or searching the web, find a drawing you wish to cutout with the Crossbow. Upon finding a suitable image you wish to cutout, you will probably need to do a bit of cleanup work on the file before you bring it into FastCAM to create the NC (Numeric

Control) code that FastCAM will use to know where to cut in, out, etc. Follow along to see how to use Gimp to clean things up a bit.

1. Select your image. The image must be black and white with crisp and clean lines if possible.

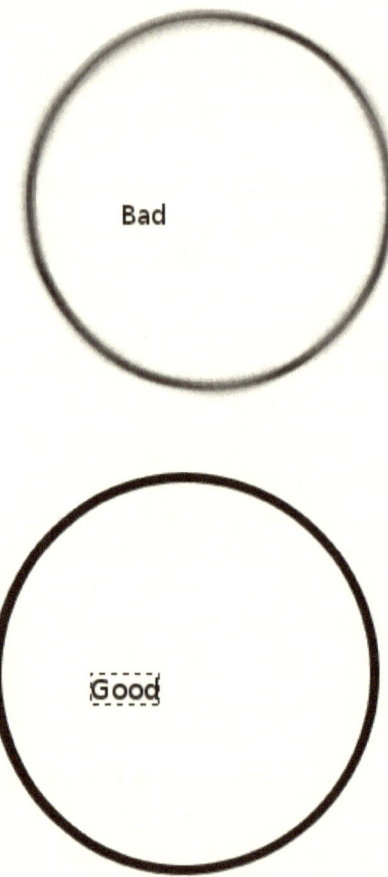

Figure 4.1 Demonstration of a clean image.

1. Open the image in Gimp and ensure the image in crisp and clean.
2. Once you are happy with the image click "File," then "Export As."

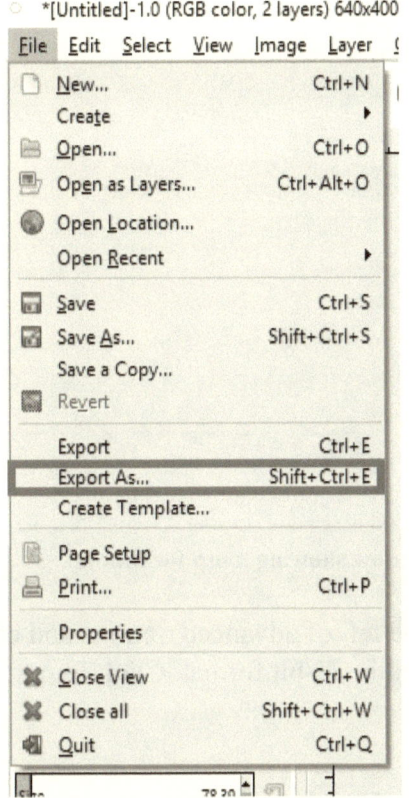

Figure 4.2 Exporting a bitmap from GIMP

3. Name the image something you'll remember and add ".bmp" to the end of the name. We'll need a bitmap for the next step. Click Export.

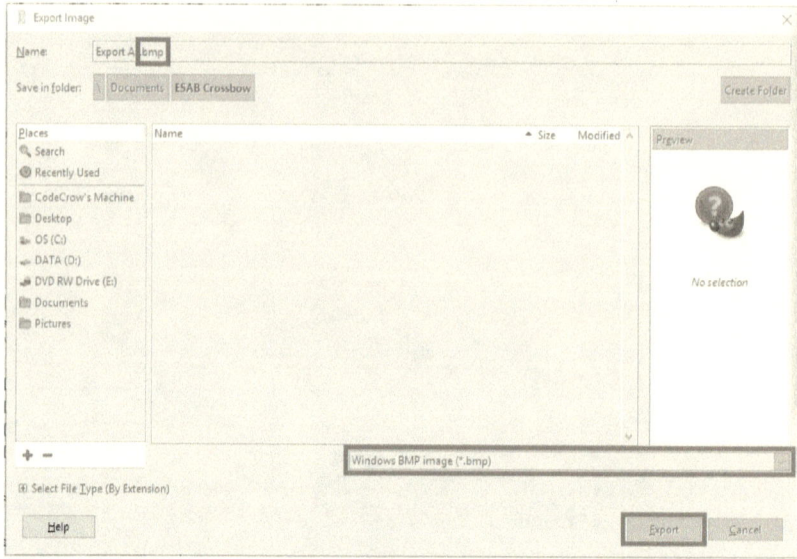

Figure 4.3 The Export window showing .bmp for bitmap.

4. Click the plus sign to the left of advanced options and ensure that the file will be saved in 24-bit format. Click Export and close Gimp.

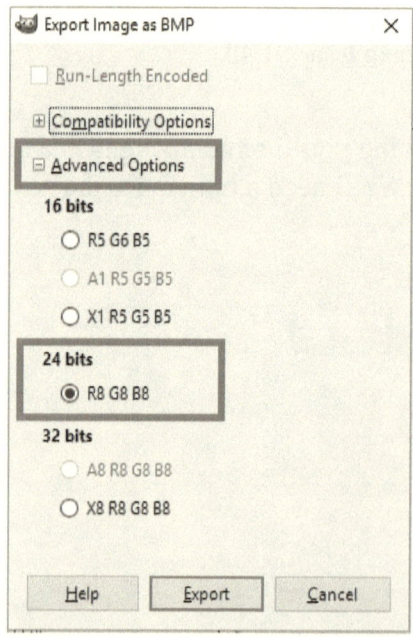

Figure 4.4 The 24-bit option is located under advanced options.

5. Open Inkscape.
6. Import the bitmap you just created with Gimp.

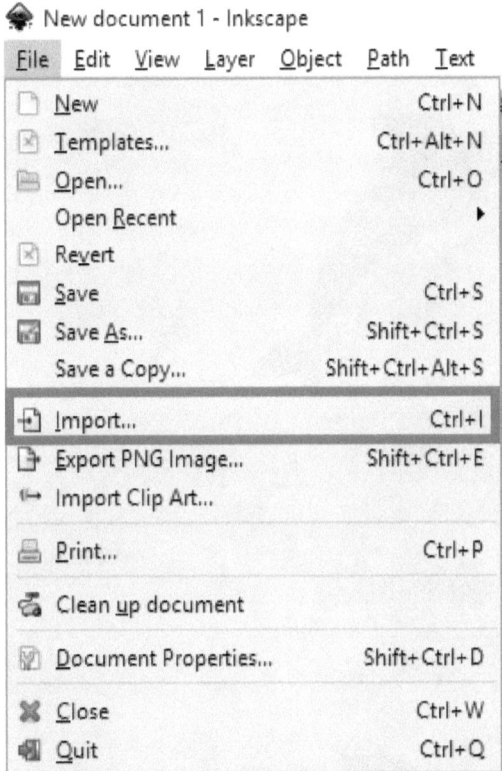

Figure 4.5 Import a file to Inkscape.

7. Click on the image after it has been imported.

Figure 4.6 Image pre-selection.

Figure 4.7 Image post-selection.

8. At the top of the screen, click Path.
9. In the drop-down menu, click Trace Bitmap.

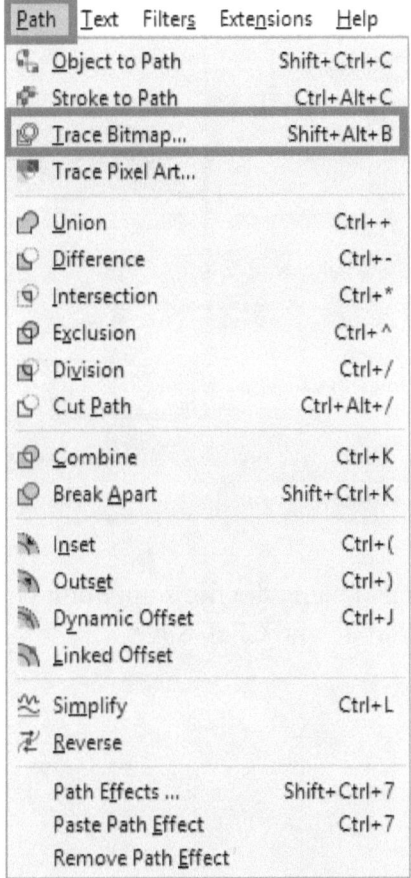

Figure 4.8 Trace bitmap function under the Path menu.

10. Click Live Preview on the right side of the pop-up interface. If the right side of the pop-up interface does not display your image, you have not selected it yet. Simply click on the image in Inkscape.
11. Tweak the settings on the left side of the pop-up interface until the image on the right looks clean. If the image was cleaned well with Gimp, this step can be skipped.
12. Once you are happy with the image on the right side of the pop-up, click Ok.

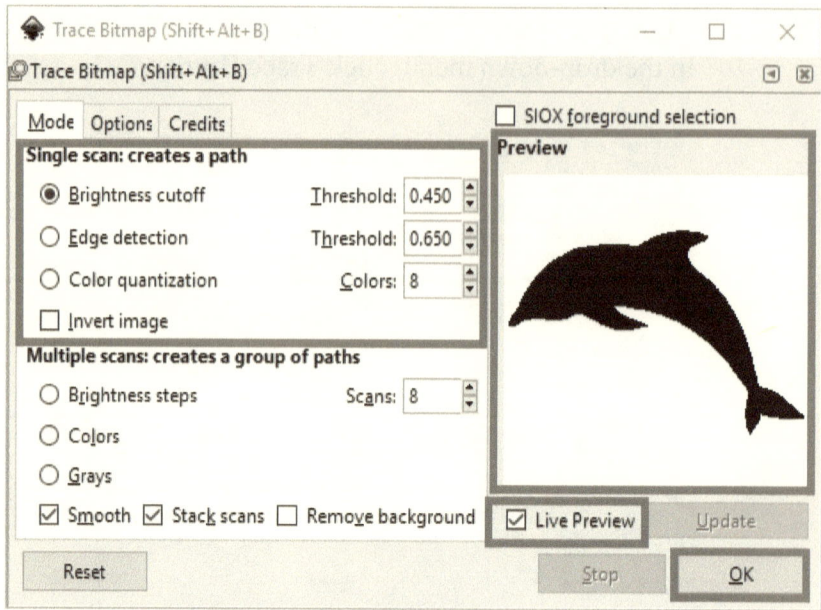

Figure 4.9 Trace Bitmaps window layout.

13. Click File, Save As and name the file something you'll remember followed by ".dxf" and Click Save.

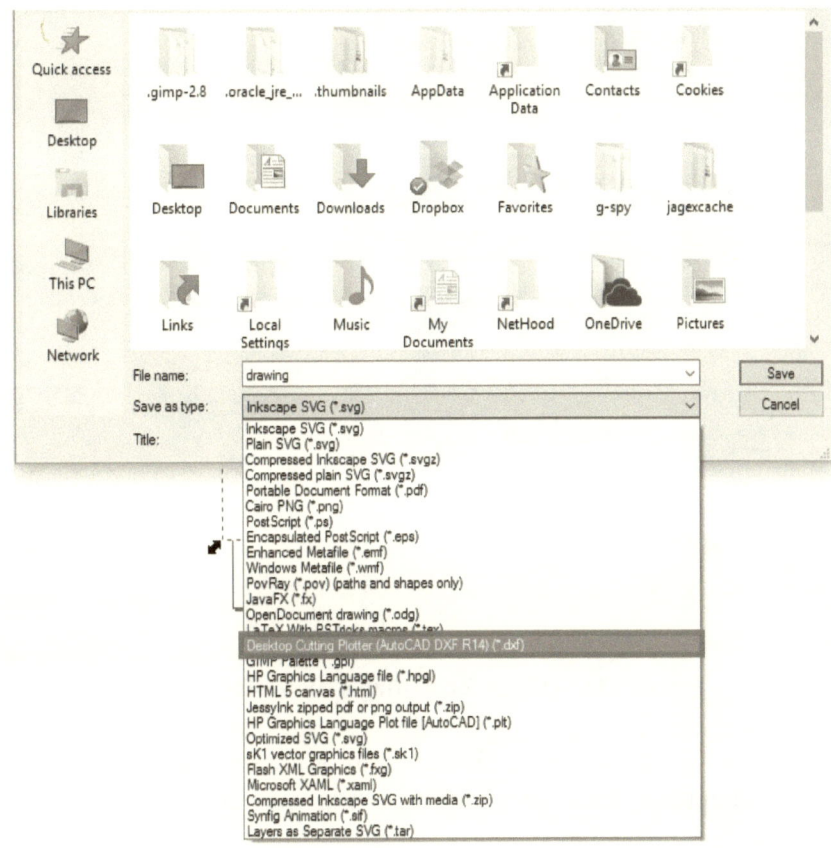

Figure 4.10 Location of dxf file extension in the save menu.

Ensure both check boxes are left unchecked and the "Base Unit" is set to metric or standard. Click Ok.

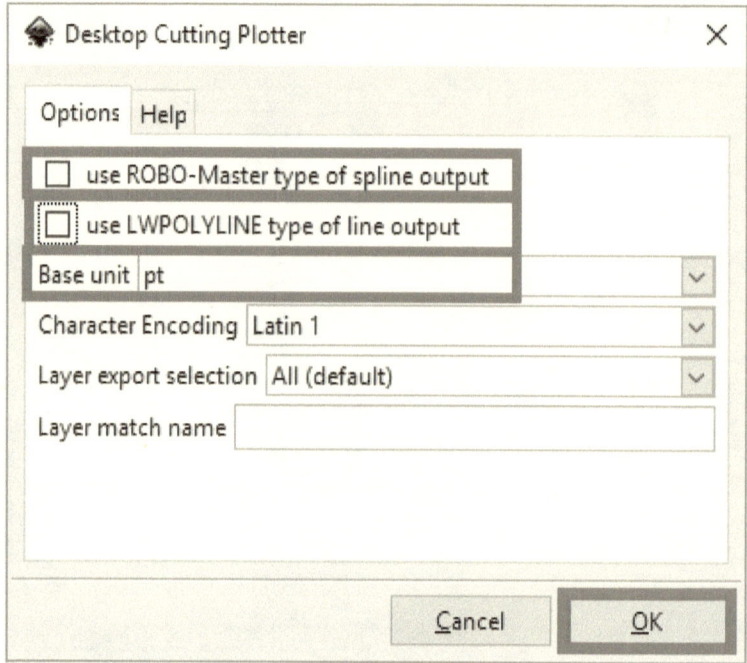

Figure 4.11 Settings for writing a successful DXF.

14. Close Inkscape. At this point you can skip ahead in this chapter to the section on Importing your drawing into FastCAM.

In the next section we will examine how to create a drawing in Inkscape. You could draw anything of course...use your creativity! What follows are steps for Inkscape version .91. Others will be similar.

Hands-on: Drawing an object in Inkscape for later use in FastCAM

* Don't be alarmed that there are quite a few steps which follow, actually you will quickly find that after a few times you will have the steps memorized!
1. Open Inkscape and click File in the top left-hand corner of the screen and select Document Properties.

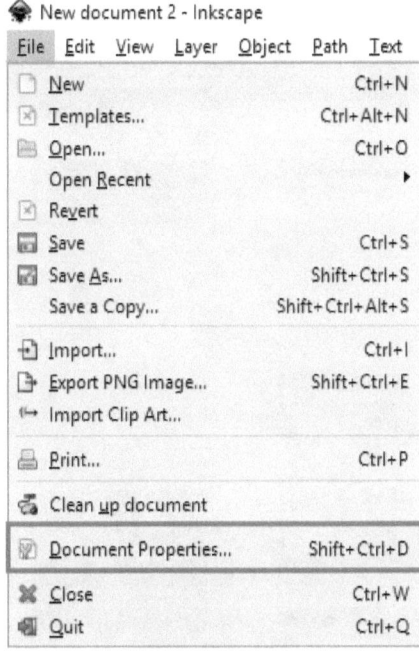

Figure 4.12

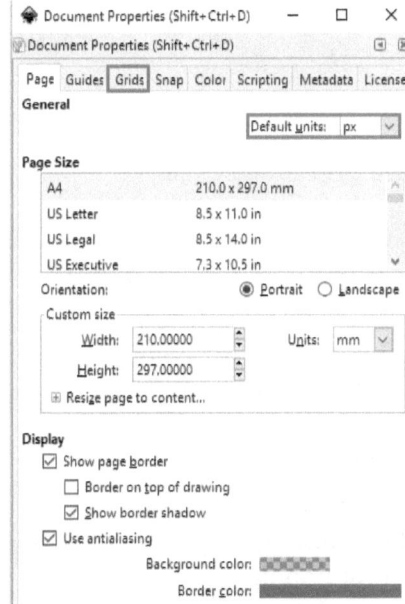

Figure 4.13

2. Change the Default Units to inches.
3. Click the Grids tab.

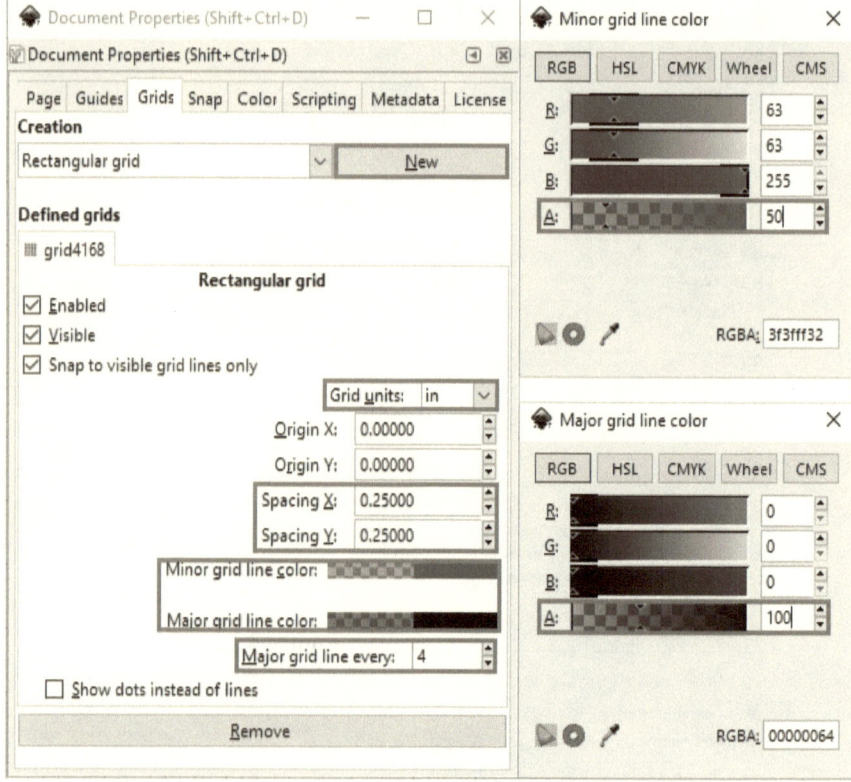

Figure 4.14

4. Remove the currently defined grid if one exist.
5. Click New to create a new grid.
6. Change the Grid units to Inches.
7. Change the Grid Spacing to 0.25 for both the X and Y sections.
8. Click on the color slider by Minor grid line color pop up.
9. Change the value by the A slider to about 50.
10. Close the Minor grid line color.
11. Click on the color slider by Major grid line color.
12. Change the color using the R, G, and B sliders.
13. Change the A slider to about 100.
14. Close the Major grid line color pop up.
15. Change the Major grid line value to 4.
16. Close the Document Properties window.

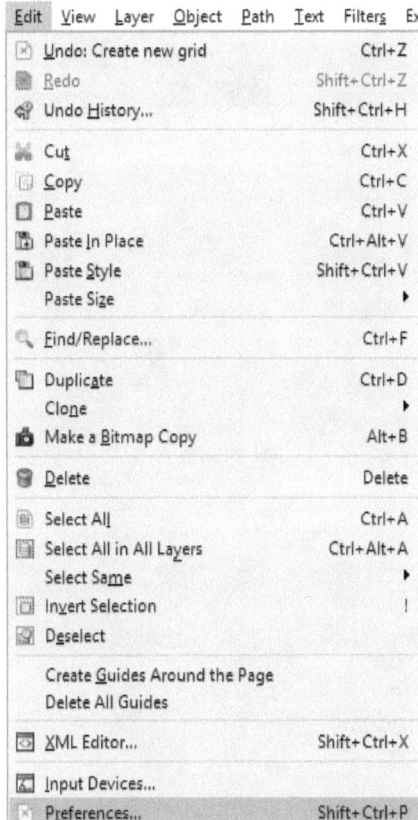

Figure 4.15

17. Click Edit in the top left-hand corner.
18. Click Preferences.

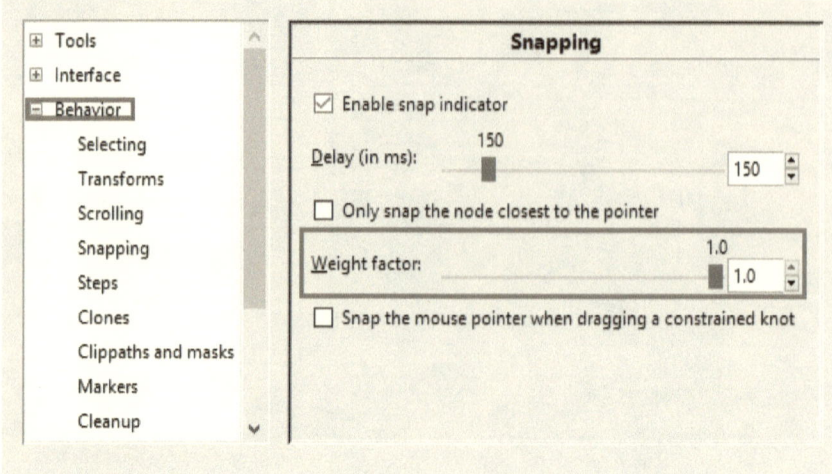

Figure 4.16

19. Click the + by Behavior.
20. Click Snapping.
21. Change the weight to 1.0.
22. Close the Preferences window.

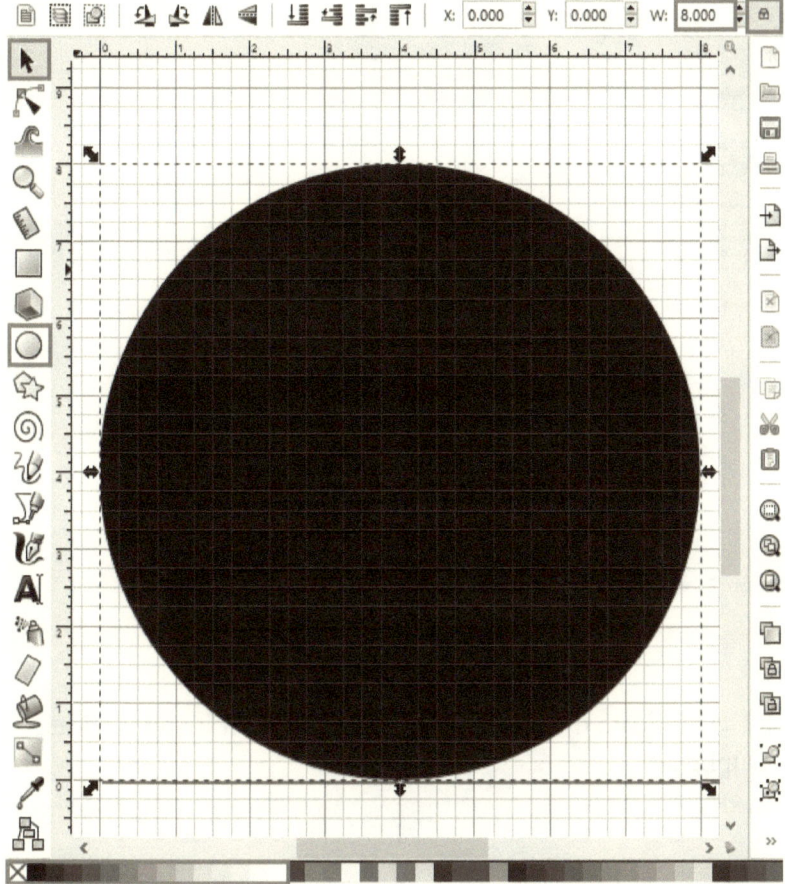

Figure 4.17

23. Zoom in by holding left control and scrolling in.
24. Click the Create Circles, eclipses, and arcs tool.
25. Click and drag to create a circle. Ensure the circle is perfectly round.
26. Click the Select and Transform objects tool.
27. Click the circle to select it.
28. Select black at the bottom of the screen.
29. Click the padlock at the top of the screen to preserve the ratio of the circle while changing its size.
30. Change the size of the circle to 8.

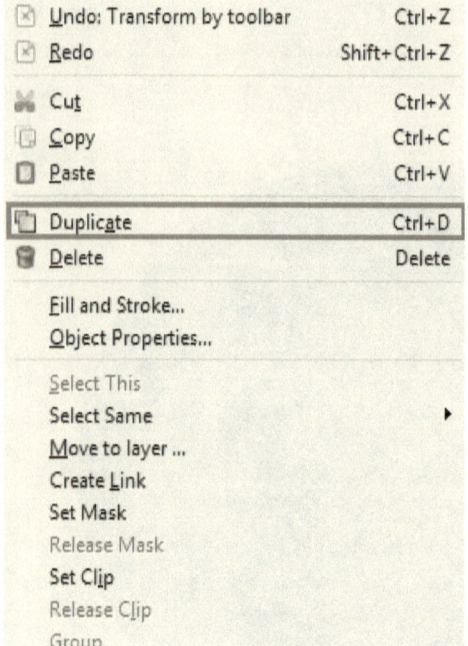

☒	Undo: Transform by toolbar	Ctrl+Z
☒	Redo	Shift+Ctrl+Z
✂	Cut	Ctrl+X
▢	Copy	Ctrl+C
▢	Paste	Ctrl+V
▢	Duplicate	Ctrl+D
▨	Delete	Delete
	Fill and Stroke...	
	Object Properties...	
	Select This	
	Select Same	▶
	Move to layer ...	
	Create Link	
	Set Mask	
	Release Mask	
	Set Clip	
	Release Clip	
	Group	

Figure 4.18

31. Duplicate the circle by right clicking on it and clicking Duplicate.

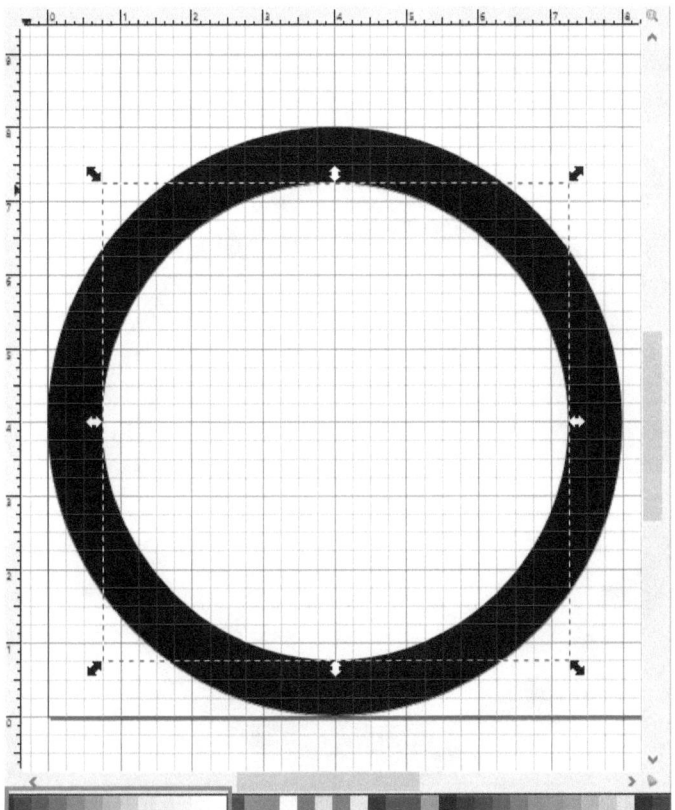
Figure 4.19

32. Click the circle to select it.
33. Select white at the bottom of the screen.
34. Ensure the new circle is selected.
35. Hold the left control key and the shift key at the same time then click and drag one of the arrows to make the circle smaller while keeping it centered inside the other circle.
36. As you drag, pay attention to the grid lines. Drag the side until you reach the third grid line. This should mean the circle is now .75 inches away from the larger circle's edge.

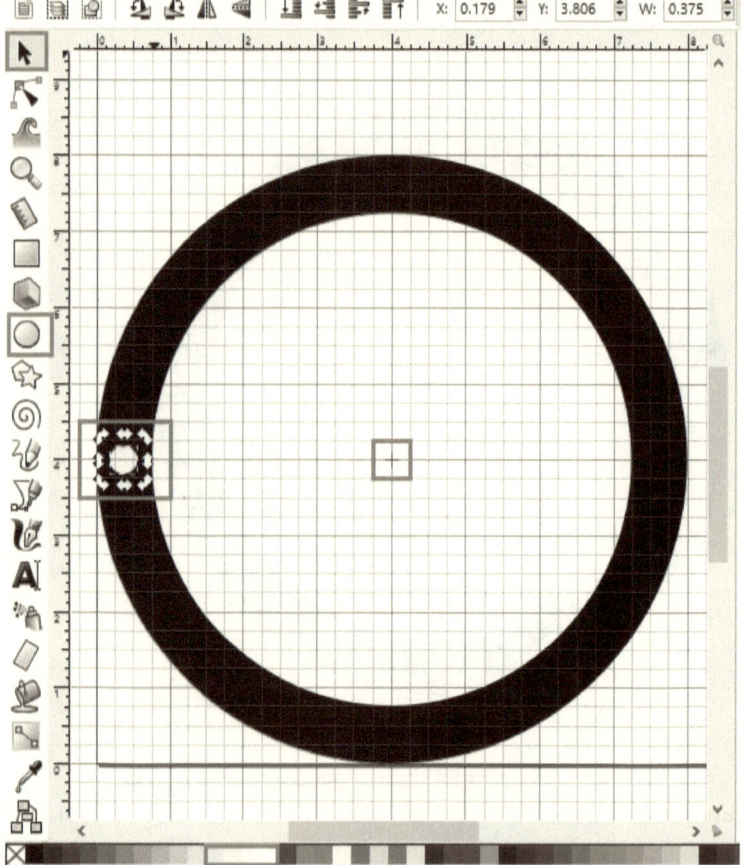

Figure 4.20

37. Click the Create Circles, eclipses, and arcs tool and click and drag to create a circle. Ensure the circle is perfectly round.
38. Click the select and transform objects tool and click the circle to select it.
39. Select white at the bottom of the screen.
40. Click the padlock at the top of the screen to preserve the ratio of the circle while changing its size.
41. Change the size of the circle to 0.375.
42. Center the new circle between the rims of the two larger circles.
43. Click the circle to select it and click the circle a second time to change into rotation mode.
44. Click and drag the crosshair in the center of the small circle to the center of the other circles.

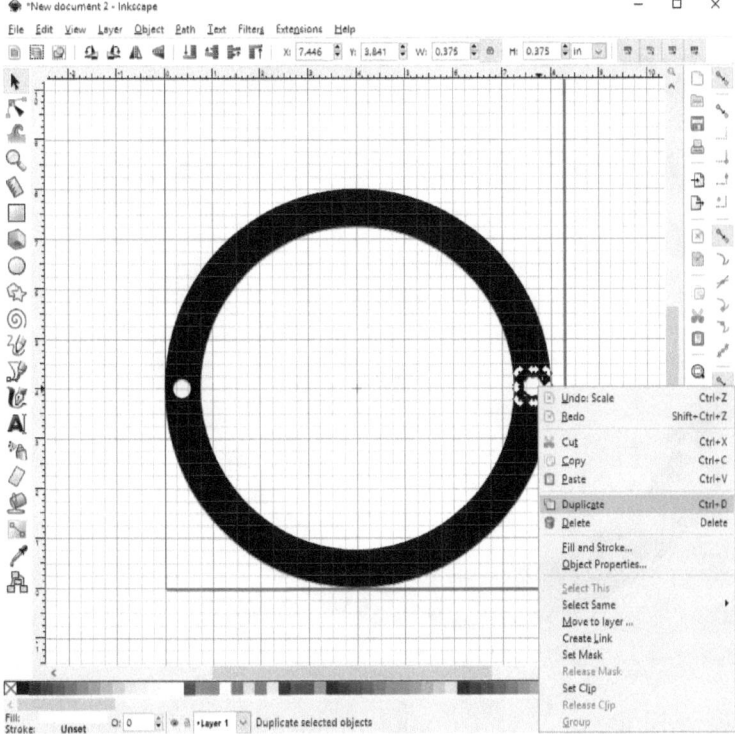

Figure 4.21

45. Duplicate the circle by right clicking on it and clicking Duplicate.

46. While holding left control, click and drag one of the handles to rotate the new circle around to the far side of the large circle.

47. Repeat steps 45 and 46 for the desired number of flange of holes.

48. When you are finished, click File and save your flange as an dxf so that you can import it to FastCAM to create your NC code for the Crossbow.

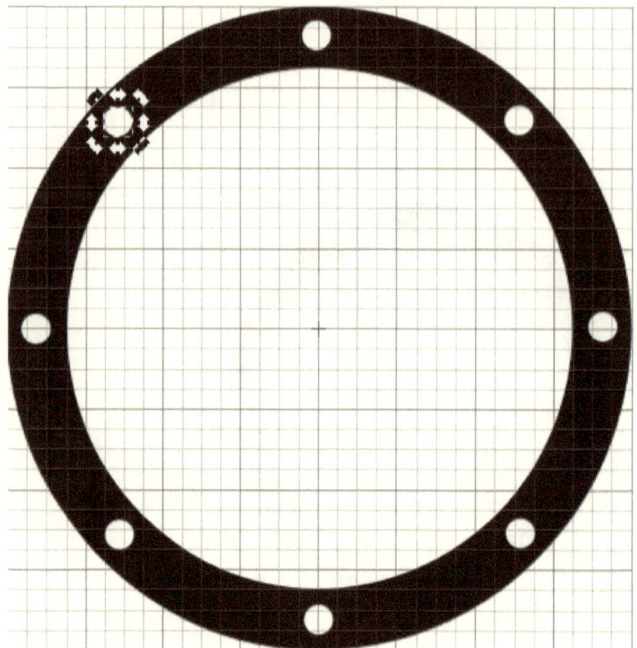

Figure 4.22 Completed design.

Next we will use a CAD (Computer Aided Design) application to create files ready for FastCAM and the Crossbow.

Hands-on: Create a drawing in LibreCAD that can be imported into FastCAM

LibreCAD is a free and open source software application which runs on PC, Mac, or Linux and easily allows you to draw 2D CAD drawings, which can be exported to DXF or IGES format which FastCAM can import to later be used to make NC g-code.

1. Step one, download and install LibreCAD from: http://librecad.org/
2. After the application is installed, start the application and select "inches" and "English".
3. Click on the Tools menu and you will see many shapes you can select. Click on the Circle, 2 Points, then click on the screen where you want a circle. Click once, then move the mouse until the desired size is obtained. Click again to create the circle. Next click the ESC key to get out of circle drawing mode.

4. Select Tools, Text, then you will see a screen where you can enter a size, font size, and text to be displayed (or cut out by FastCAM). Change the text to sometime you wish to display, and change the size to 6. Select the cursive font for now. Experiment with other sizes and fonts.

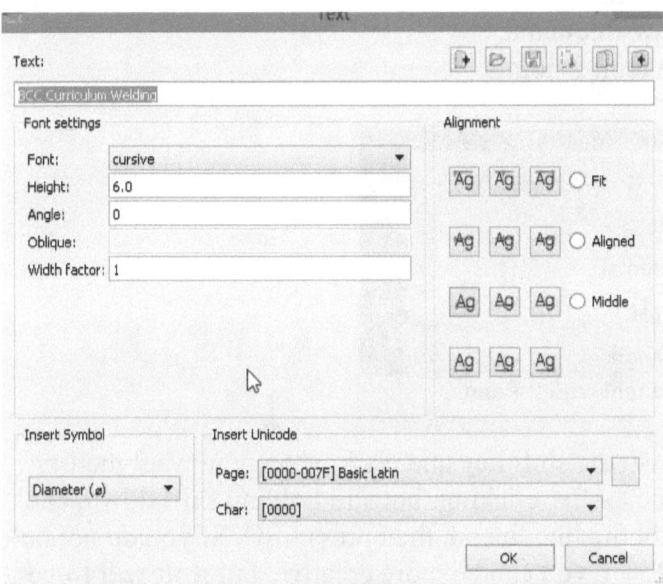

Figure 4.23 Text settings LibreCAD.

5. Move the cross hairs where you want the text, click the left mouse button, then press ESC to escape text mode.

Figure 4.24 Output of text in LibraCAD.

You do not need to save your work at this point, this last part with text and circles were for practice. Next try to create an object which we will export to the Crossbow, in this case a bracket.

Hands-on: Create an shape and output to FastCAM
1. Start LibreCAD from scratch. To draw, you have built-in shapes or you can draw free-hand.
Select the line tool, 2 points.

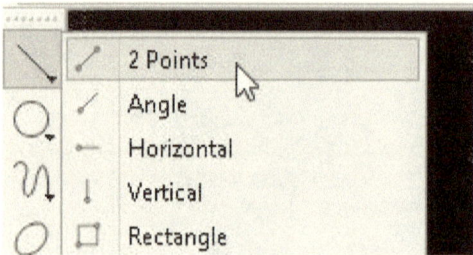
Figure 4.25 Line tool with 2 Points.

Then, hold down the shift key and click where you want the first point of the line. After clicking, but still holding shift down, click where you want the line to end, then press ESC. If you do not hold the shift key down, you can be more creative, but shift will force it to use 45 degree increments, which may help your design as you could otherwise get a crooked line.

2. Continue to draw lines until you have the following bracket.

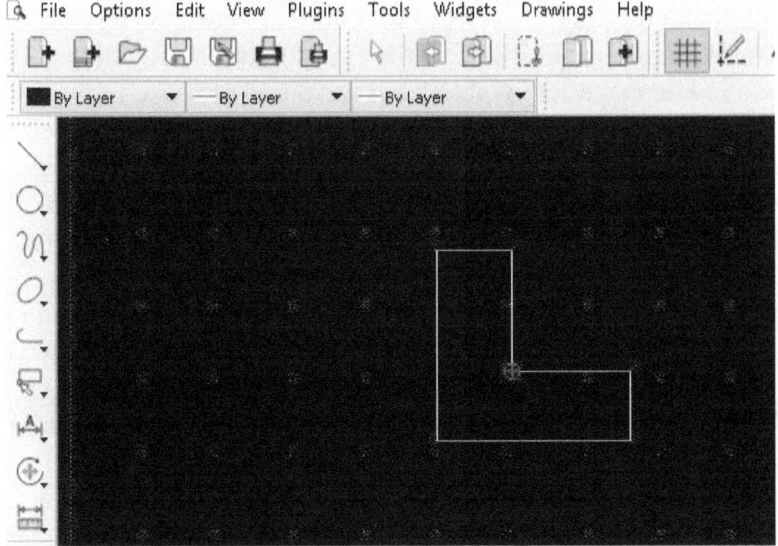
Figure 4.26 Draw this bracket.

3. You may want to select View, Zoom In (and Zoom Out) to check that your lines all touch and are connected.

4. Save your work to DXF format by selecting File, Save As, then select the location to save to, next change the format to DXF R12 and click Save. Make sure you save to a location you can access! You can now close LibreCAD.

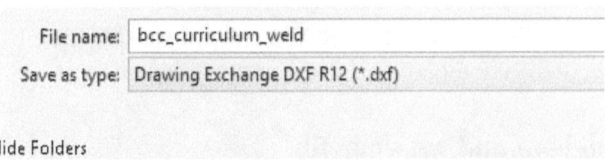

lide Folders
Figure 4.27 Output to R12 DXF format from LibreCAD.

5. To import to FastCAM, after making sure the hardware key is installed in the computer, start FastCAM then select Files, DXF Restore, then select Inch, Enter, and browse to where you saved the DXF file in step 4.

6. Click on the file and press Select.

7. With luck you can now select Program Path, FastPATH, Start FastPATH, then Start FastPATH, then yes to output NC code. Give it the drive letter of a removable flash drive you can take to the Crossbow, and a name you will remember. Try not to use spaces in the name, although underscores work. E.g. my_file_bracket3.txt
* if it will not outline, or if there are many "entities" noted in the upper left, you may have issues. Many entities would be over 400 or so.
8. Open the file in a text editor, such as Wordpad or Notepad and delete the first four lines or so of comments. You should only have lines starting with G and a number such as "G31". This is the "G code" which are directions for location of the cutter head, on and off, etc. To see them, visit this link:
http://www.machinemate.com/FullListCodes.htm

Importing your drawing into FastCAM
Since FastCAM is a physical key-based system, you will need the FastCAM serial USB key inserted into a computer with FastCAM installed. You also will need a second USB flash drive to save your drawing to for use in the EASB Crossbow. Have this ready before you start!

Using the steps listed in Chapter 3 of this text, you already performed a DXF restore, and set appropriate cut in, cut path, and cut out points then create the NC g-code file for use in the Crossbow, so the steps previously mentioned should not be new to you.

Hands-on: How to load and cut your file
Once your file is exported from FastCAM for use on the Crossbow(c) you are ready to load your file and run a test pass, then start cutting. Follow these steps:

1. Turn on the air compressor, adjust air pressure on it.
2. Make sure the Crossbow is off.
3. Insert the flash drive.
4. Turn on the Crossbow, *making sure the red emergency top button is out and not depressed* before you turn the Crossbow on. If it is not you will get to a point where "Man w<->h" will

flash in the center bottom of the screen and the final design to be cut out will not display.

5. Press F3 to open the Edit Menu.
6. Press F6 to open the USB Menu.

Figure 4.28 USB Menu.

7. Press F1 to load a file from the USB.

Figure 4.29 Load menu.

8. Using the arrow keys, navigate to your file slowly.
9. Press F7 to preview the file.

Figure 4.30 Selection buttons.

10. Press F7, which is "Next."

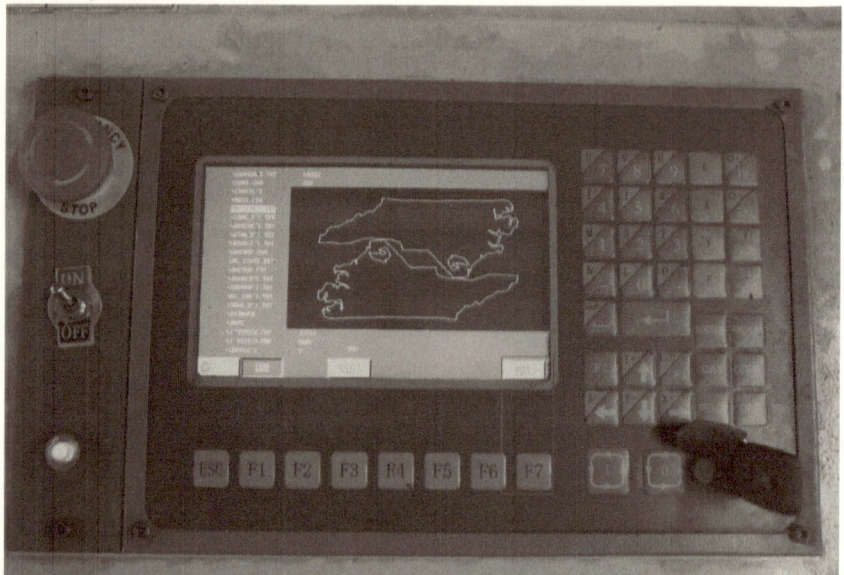

Figure 4.31 Previewed file.

11. Press Esc twice to return to the Main Menu.

Figure 4.32 Escape button.

12. Press F1 to enter Auto Mode.

Figure 4.33 Auto Mode button.

13. Press F4 to preview and load the drawing into the cutting mode.

Figure 4.34 Preview button.

14. (Optional Step) Press F7 to view additional information on the file. If F7 was pressed, press the Esc key after you have finished viewing the additional information.

15. Press Esc to return to the cutting menu.

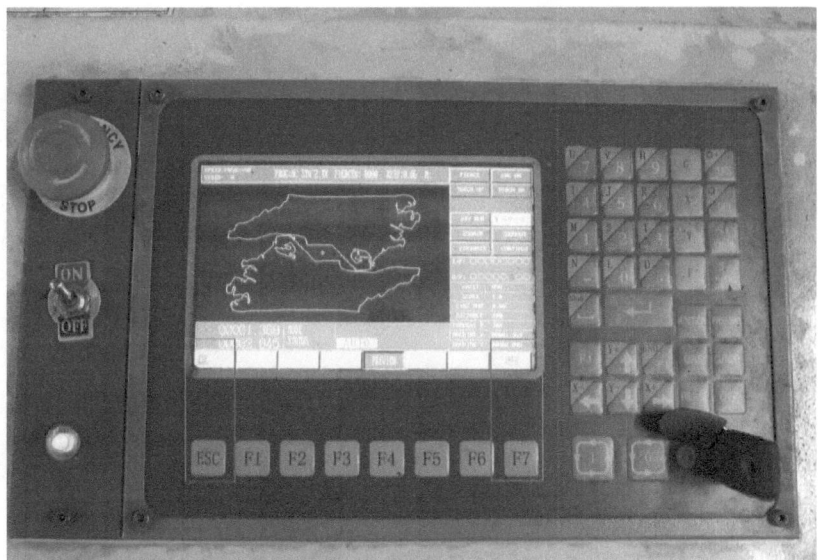

Figure 4.35 Escape and Information buttons.

16. Press F6 for more options.

Figure 4.36 More button.

17. Turn off the Crossbow's motors via the toggle switch located at the top-left of the control panel.

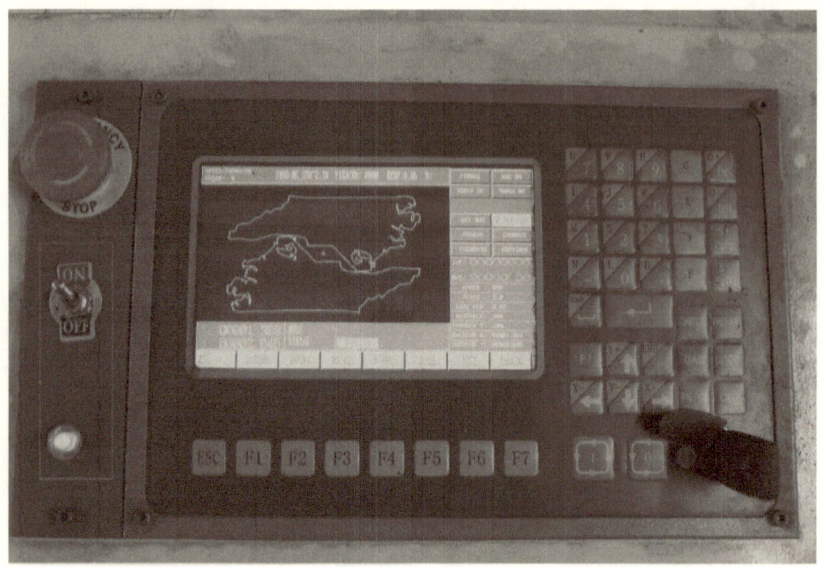

Figure 4.37 Motor power switch.

18. Position the cutter's head in the appropriate starting point slowly.

19. Press F1 to outline the cutting area.

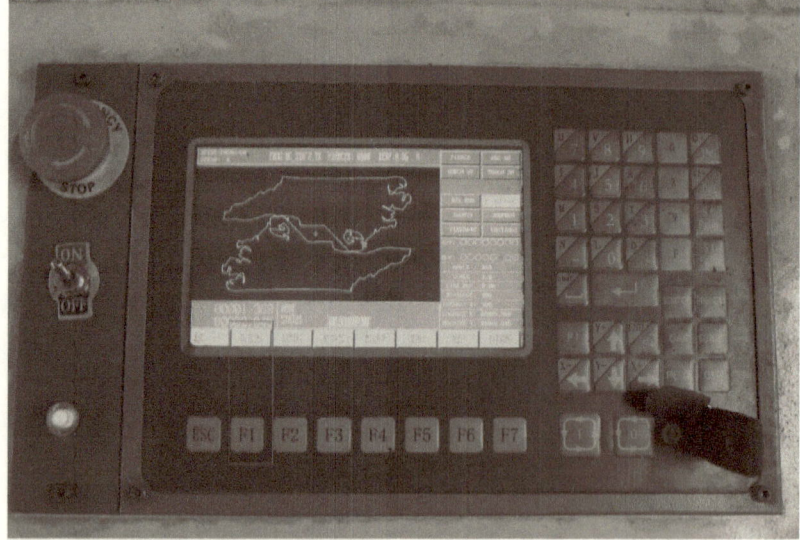

Figure 4.38 Outline cutting area button.

20. Select the starting point when prompted (Recommended starting point is "Start From Here.")

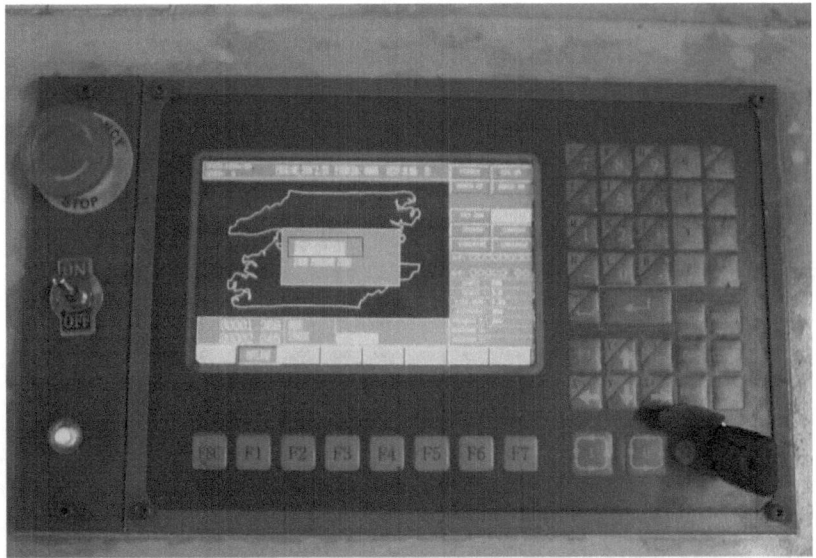

Figure 4.39 Enter Button to begin cutting.

21. Reposition the head repeating steps 17 – 19 until you are happy with the placement.

22. ![note icon] Optional but very important step: Press X to toggle a dry run. This will trace the file, but will not cut. This will also not return to its starting point.

23. Turn on the plasma cutter and check that air is on and pressure is good on the plasma cutter.

24. Press the green start button.

25. When finished, make sure the Crossbow, plasma torch, and the air compressor are all shut off if no longer in use.

26. Remove the flash drive.

This text has covered the basics of creating custom designs for the *Crossbow*. Practice makes perfect (or so they say) so try to hone your skills with the following challenges.

Of particular interest is the first link below under **Links**, which is an *Inkscape* DXF export plugin which might help you create better export files from *Inkscape*. Add it to your *Inkscape* install and experiment as you expand your skills. Where to go from here? Some things you should look into would include:

a) Use Google Sketchup to draw and export to DXF.
https://www.youtube.com/watch?v=ZigQyJ9nfFU

b) Use a image scanner to scan a picture to DXF format (after cleanup in Gimp, Inkscape, or other applications.

c) Explore some of the many other free/open source image programs available via a Google search such as "Paint.NET" and others.

d) Look into using open-source milling tools and programming languages such as PyCAM and others.
https://www.youtube.com/watch?v=stcZAIy2xnE
http://www.craftsmanspace.com/free-software/free-cam-software.html

Reinforce your skills
Try the following to reinforce what you have learned.

1) Using LibreCAD, create a drawing you could cut out which has circles.

2) Create a drawing you could cut out which could be used as a hinge, bracket, or some form of support brace. You might make some interesting decorative hinges, then weld/bend the final parts for use with a pin.

3) Create an outlined word or initials that could be cutout with the Crossbow.

4) Create cutouts which would form three or four sides of an object. E.g. if you were making a square lantern which held a candle perhaps. Draw the sides, top, base, etc. and cut out the parts, then weld them together to create your candle lantern. Google for design ideas.

5) Create a kool retro-art "tiki" or other art-themed cutout.

6) Create a Halloween, Christmas, Easter, nautical, hunting, hotrod, or other themed cutout.

7) Create a decorative addition to a mail box, wall, or yard art. E.g. a chicken outline for a farm, an Easter bunny you could tack a metal rod to for yard decoration, a cutout of a logo or image (say a rifle, palm tree, religious symbol, or other) for decoration.

7) Find a URL for a site not listed in this chapter which would help you with the Crossbow or operations listed in this chapter.

Links
http://www.bobcookdev.com/inkscape/inkscape-dxf.html
http://www.machinemate.com/FullListCodes.htm
https://openclipart.org/unlimited-commercial-use-clipart
http://www.readytocut.com/community/forums/cnc-art-file-sharing.136/

Notes:

_____
_____
_____

_____
_____
_____
_____
_____
_____
_____
_____
_____
_____